Maple V Release 3 for DOS and Windows

Getting Started

Brooks/Cole Publishing Company

An International Thomson Publishing Company

Pacific Grove Albany Bonn Boston Cincinnati Detroit London Madrid
Melbourne Mexico City New York Paris San Francisco Singapore Tokyo
Toronto Washington

Maple V Release 3 for DOS/Windows: Getting Started © 1995 by Waterloo Maple Software. All rights reserved. No part of this work may be reproduced, stored in a database or retrieval system, or transcribed, in any form or by any means—electronic, mechanical, photocopying, or otherwise—without the prior written permission of the publisher, Brooks/Cole Publishing Company, Pacific Grove, California 93950.

Maple is a registered trademark of Waterloo Maple Software. All other product names mentioned in this work are trademarks or registered trademarks of their respective companies.

Printed in the United States of America

10 9 8 7 6 5

For more information, contact: Brooks/Cole Publishing Company
511 Forest Lodge Road
Pacific Grove, CA 93950   USA

For orders, call: 800-354-9706 (8:00 A.M.–5:00 P.M. Eastern Time)
For technical support:
    call: 800-214-2661 (6:00 A.M.–4:00 P.M. Pacific Time)
    fax: 408-375-0120
    email: SUPPORT@BROOKSCOLE.COM

International Thomson Publishing
Berkshire House 168–173
High Holborn
London WC1V 7AA
England

International Thomson Publishing GmbH
Königwinterer Strasse 418
53227 Bonn
Germany

Thomas Nelson Australia
102 Dodds Street
South Melbourne, 3205
Victoria, Australia

International Thomson Publishing–Asia
221 Henderson Road #05–10
Henderson Building
Singapore 0315

Nelson Canada
1120 Birchmount Road
Scarborough, Ontario
Canada M1K 5G4

International Thomson Publishing–Japan
Kyowa Building, 3F
2-2-1 Hirakawacho
Chiyoda-ku, 102 Tokyo
Japan

Contents

Introduction 1
   How to Use This Guide 1

Installing Maple V Release 3 3
   Objectives of This Chapter 3
   Contents of This Package 3
   If You Run into Problems 3
   Installation 4
      Floating-point Coprocessors and Maple 4
      Disk Space Requirements 4
   How Maple V Uses The Hard Disk 4
      Installation Procedure 5
      Setting Up for Microsoft Windows 6

Using Maple V Release 3 for Windows 7
   Objectives of This Chapter 7
   Starting Maple V 7
   Exiting Maple V 7
   Maple Commands versus Interface Commands 8
   Interrupting a Maple Calculation 8
   The MAPLE.INI File 8
   Increasing Maple's Memory Allocation 8
   Maple's Internal State 9
      Saving Maple's Internal State 9
   Types of Maple V Windows 10
      Worksheet Window 10
      Help Window 10
      Two-dimensional Plot Window 11
      Three-dimensional Plot Window 11
      Animation Window 11

Contents

Working with Worksheets 11
Worksheet Regions 12
    Input Regions 13
    Text Regions 13
    Output Regions 14
    Graphics Regions 14
    Region Groups 14
    Editing Worksheets 15
    Recalculating a Worksheet 16
Converting Maple V Worksheets 16
    Between Release 2 and Release 3 16
    Among Release 3 Platforms 17
Manipulating Plots 17
    Two-dimensional Plots 18
    Three-dimensional Plots 20
    Animation 22
Maple V Menus 23
    File Menu 23
    Edit Menu 24
    Format Menu 25
    View Menu 26
    Options Menu 27
    Help Menu 27
Maple V's Help Facility 28

Using Maple V Release 3 for DOS 31
Objectives of This Chapter 31
Starting Maple V 32
Exiting Maple V 33
Interrupting a Maple Computation 33
Maple Initialization Files 33
    Multiple Initialization Files 35
Maple Environment Variables 36
    MAPLECOL 36
    VIDEOMD 37

FOREGR=XXXXXX and BACKGR=XXXXXX	38
MAPLEPRT=type	38
MAPLEPRF=name	38
SESSMAX=n	38
Compatibility Considerations	38
Virtual Memory	39
Maple V Memory Requirements	40
Using Maple V on a Network	41
The Status Line	41
Editing in Maple V	42
The Command Line Editor	42
Expression Editing	42
File Editing	43
Session Review Mode	44
Input/Output Capture Mode	44
Printing the Session Log	44
Preserving Maple's State	44
Manipulating Graphical Output	45
Two-dimensional Plots	45
Three-dimensional Plots	46
Animation	48
Printing and Saving Plot Output	48
Supported Printer Device Types	49
Using Maple V Plots in Other Programs	50
Using the Maple V Menus	50
Level One	51
Level Two	52
Level Three	52
A Sample Menu Entry	53
Maple V's Help Facility	54

Maple V Release 3 for DOS Running Under Windows   57

Setting Up Maple V under Windows	57
Running Maple V under Windows	57
Special Features for Windows	57

Contents

    Copy and Paste 57
    Multiple Maple V Sessions 58

Other Applications 59
Mint: The Syntax Checker 59
March: The Maple Archive Manager 59
Mapledit: The Maple Editor 61
m2src 61
updtsrc 62
Multiple Library Access 62

Introduction

This guide supplements the *Maple V Flight Manual* and the *Maple* help pages by providing a description of the details specific to Microsoft Windows and DOS users. The chapter *Installing Maple V Release 3* describes the installation procedure and how *Maple V* makes use of your system's resources. The chapters titled *Using Maple V Release 3 for Windows* on page 7 and *Using Maple V Release 3 for DOS* on page 31 describe, respectively, the Windows and DOS interfaces for *Maple V*.

How to Use This Guide

This guide is intended to be used by anyone wishing to install or use *Maple V Release 3* for DOS and Windows. Throughout this document, different typefaces or type styles are used to indicate information of importance:

Italics are used to indicate reference material. This reference material are either other *Maple V* books or guides that you can consult or other sections within this document.

Boldface is used to indicate keystrokes that you are required to type or menu items that you are required to choose. When a keystroke includes a + sign, as in **Shift + Return**, you should hold down the first key while typing the second key. **Boldface** is also used to indicate labels or information that you see on the screen. For example, we would refer to the Windows **Program Manager** in boldface.

The **helvetica** font is used to indicate standard *Maple* commands—for example **quit**, **plot3d**, **animate**, and **int**.

Installing Maple V Release 3

Objectives of This Chapter

This chapter describes the procedures for installing the *Maple V* Computer Algebra System on your computer running PC-DOS or MS-DOS version 3.3 or higher or Microsoft Windows version 3.1. Before beginning the installation, check that your computer fulfills the following requirements:

- Any 80386- or 80486-based computer
- 12–15 megabytes (MB) of free hard disk space (an additional 3 MB is needed during installation)
- At least 4 megabytes of RAM
- A high-density disk drive
- Optionally, a 80387 or 80487 numeric coprocessor
- Microsoft Windows version 3.1 is required to run the Windows version.

These are the minimum requirements for installing and running *Maple V Release 3*. If your computer has more memory, *Maple V* can make use of it. Please read the entire installation section carefully before proceeding with the installation process.

Contents of This Package

Your *Maple V Release 3* Student Edition installation package contains four 3.5-inch installation disks, this *Maple V Release 3 Getting Started* booklet, and a *Release 3 Notes* booklet.

If You Run into Problems

If you have any questions about the installation procedure, or encounter any technical problems, please call the Technical Support Department at Brooks/Cole Publishing

Installation

Company at (800) 214-2661, weekdays between 6:00 AM and 4:00 PM (Pacific Time). We can also be reached by FAX at (408) 375-0120 or by electronic mail at SUPPORT@BROOKSCOLE.COM

Installation

Floating-point Coprocessors and Maple

Maple V Release 3 runs on PCs that have an 80387 or 80487 coprocessor and on PCs that do not have a coprocessor installed. The most notable difference in performance is that *Maple V* performs numerical calculations significantly faster on a machine with a coprocessor. As a result, a machine that has a coprocessor installed produces plots much faster. The speed with which *Maple V* performs symbolic calculations is identical in both cases.

Disk Space Requirements

Depending on what installation configuration you choose, *Maple V Release 3* requires between **12 MB** (megabytes) and **18 MB** of disk space during installation. Before installing *Maple V Release 3*, make sure there is sufficient free space on your hard disk to contain all the files. If you use disk compression software, then you may need up to 3 MB more disk space during installation than is finally occupied by *Maple V*.

How Maple V Uses The Hard Disk

Maple V can be installed anywhere on your hard disk. When installing *Maple V*, you will be asked on which disk and in which directory you wish to install it. The default is **C:\MAPLEV3**. We subsequently refer to the directory in which you install *Maple V* as the *Maple* directory. Examples will use **C:\MAPLEV3**, so you will have to make the appropriate substitutions when following the examples if you have installed *Maple V* in a different directory.

Under the *Maple* directory, the installation program creates several subdirectories. The two most important subdirectories are

- **BIN** The directory in which the executable programs that comprise *Maple V* are stored with the system initialization files. This directory should appear in a DOS **PATH** statement in your **AUTOEXEC.BAT** file. For example:

    PATH C:\;C:\DOS;C:\MAPLEV3\BIN

Installing Maple V Release 3

This line will be automatically inserted by the installation procedure.

- **LIB**   The directory that contains the *Maple* library and menu files. *Maple V* assumes that this directory is **\MAPLEV3\LIB** on the current disk. This can be overridden by setting a DOS environment variable called **MAPLELIB** to the appropriate directory name. For example, if you installed *Maple V* on drive **E:** in the directory **\APPS\MAPLE**, the following statement should appear in your **AUTOEXEC.BAT** file:

    SET MAPLELIB=E:\APPS\MAPLE\LIB

    This line will be automatically inserted by the installation procedure.

When *Maple V* creates temporary files, as it does when creating plots or displaying a help page, these files are created in the current directory and are erased as soon as *Maple V* is finished with them.

Installation Procedure

1. Insert the diskette labeled *Maple V Installation 1* in drive **A:** or **B:**

2. If you are installing *Maple V* on a machine that does not have Windows, change to the drive in which you placed the diskette, and at the DOS prompt, type

    DOSINST

    to start the *Maple V* installer.

    If you are installing *Maple V* on a machine that does have Windows, start Windows, select **Run** from the **File** menu of the Windows **Program Manager**, and enter

    A:\INSTALL

    to start the *Maple V* installer. If you placed the diskette in drive **B:** then change the above command appropriately.

3. You will be shown a dialog that allows you to customize your *Maple V* installation.

    If you are installing under DOS, you will be unable to install the **Windows Maple** version or the **Examples** directory.

    When you install only the DOS version of *Maple V*, the installation takes more than **12 MB** of disk space. When you install both the Windows and DOS versions, the DOS version takes up much less space because both versions use the same library files and structures.

Getting Started

4. When you have selected the components to be installed, you are presented with two other dialogs. First, the **Installing File:** dialog informs you of the progress of the installation. Also, from time to time, another dialog appears, asking you to switch diskettes.

5. At the end of the installation, if you have installed the DOS version of *Maple V*, you are prompted whether to make changes automatically to your **autoexec.bat** file. We recommend that you select **yes**.

A full installation of *Maple V Release 3* requires approximately **18 MB** of disk space. This includes the *Maple* Library, the **Mint** and **March** applications, the DOS version of *Maple V Release 3*, the Windows version of *Maple V Release 3*, a part of the Share Library, the Examples directory, and the utilities **updtsrc** and **m2src**. *Maple V Release 3* will use less hard disk space if you do not install certain components.

If you have installed the DOS version, reboot your computer after the installation.

Setting Up for Microsoft Windows

If you have installed the Windows version of *Maple V Release 3*, several Windows icons and a program group will be set up automatically, including:

- A **Maple V Release 3** program group

- A **Maple V Release 3 for Windows** icon

- A **Maple V Release 3 for DOS** icon

- A **Mint** application icon

- A **Maple Text Editor** application icon

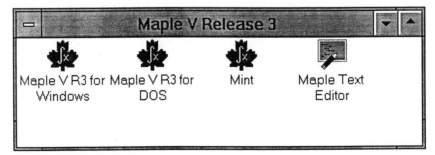

When the installation is complete, you should have a **Program Group** window that looks similar to the one shown here.

Using Maple V Release 3 for Windows

Objectives of This Chapter

This chapter describes the user interface features of the Windows version of *Maple V Release 3*. It assumes that *Maple V* is correctly installed and running under Windows 3.1 on your computer. To install *Maple V* on your system, refer to the chapter *Installing Maple V Release 3* on page 3. It also assumes that you are familiar with the standard Windows interface and know the meaning of basic terms such as *mouse button*, *double-click*, and *window*.

The purpose of this chapter is to familiarize you with the Windows-specific aspects of *Maple V*'s user interface. Supplementary information is contained in the *Maple* help pages. For information on how to use *Maple* commands or the *Maple* programming language, refer to the *Maple V Flight Manual* (by Ellis, Johnson, Lodi, and Schwalbe) and *Maple V Quick Reference* (by Blachman and Mossinghoff) and the help pages.

The Windows version of *Maple V* offers an interface with features designed to conform to the standard Windows interface, allowing you to create worksheet documents that combine input, output, text, and graphics.

Starting Maple V

To start *Maple V* for Windows, double-click the **Maple V Release 3 for Windows** application icon. This action opens the application with a blank worksheet.

Exiting Maple V

To exit *Maple V*, choose the **Exit** command in the **File** menu or enter **Alt + F4**. If the current *Maple V* document has been changed and these changes have not been saved, *Maple V* prompts you to save changes before exiting.

You can also exit *Maple V* by entering the **quit**, **done**, or **stop** command. No opportunity to save your work is given, and *Maple V* is exited immediately.

Maple Commands versus Interface Commands

Throughout this guide the term *command* is used in two different contexts. It is important to distinguish between the two.

Maple commands are those that are sent through input regions to *Maple*'s mathematical engine. The mathematical engine, or kernel, performs all the calculations in a session and is the foundation on which the *Maple* programming language is built.

Interface commands represent actions, or transformations, made on a worksheet and do not affect mathematical calculations. These commands are accessible through menu choices or keystroke combinations. The term *menu item* may also be used to indicate interface commands.

Interrupting a Maple Calculation

To interrupt *Maple* during a long calculation, press **Ctrl+Break** or click the **STOP** button on the Tool Bar in the worksheet window. The *Maple* prompt will usually appear within a few seconds. If you have sent a long sequence of commands to *Maple* (for example, via *Maple*'s **read** command), interrupting *Maple* halts calculation of the current command and prevents the execution of any further commands in that sequence. **Note:** *Maple*'s floating-point operations will not be interrupted.

The MAPLE.INI File

You can set up *Maple* to execute automatically a series of commands whenever it is started. To do this, use a text editor (for example, the *Maple* text editor) to put the desired commands into a text file named **MAPLE.INI** located in the current working directory (that is, **C:***backslash***MAPLEV3***backslash***BIN**, unless you have specified another working directory).

This **MAPLE.INI** file is especially useful for setting *Maple* global variables such as **Digits** and **Order**.

Note: Make sure that this file ends with a return character.

Increasing Maple's Memory Allocation

Maple V is configured to use as much free memory as is available at any given time.

Using Maple V Release 3 for Windows

If you wish to increase the amount of memory available to *Maple* and you have some free disk space available, select the **386 Enhanced** icon in the Windows 3.1 **Control Panel** and increase the amount of virtual memory configured for Windows. **Note:** Increasing Windows' virtual memory does require disk space. However, we do recommend that you increase the virtual memory size if you have the free disk space and want to do memory-intensive calculations with *Maple*.

If you do run out of memory while using *Maple V* for Windows, you will get a message box indicating that you are out of memory. *Maple* will give you an opportunity to save your work.

Maple's Internal State

An important distinction must be made between the state of a worksheet and *Maple*'s internal, or kernel, state. Internally, *Maple* stores only the most recent value defined for a variable in a session. For example, if the variable x is initially defined to one value and later in the session is redefined, *Maple* remembers only the most recently defined value. *Maple*'s mathematical engine does not remember previous values assigned to a variable.

Never assume that you can determine *Maple*'s internal state merely by looking at the state of the worksheet. If you have any doubts about the value of a certain variable, enter the variable name as input followed by a semicolon and the value will be displayed.

Saving Maple's Internal State

In the **File** menu, the **Save** and **Save As...** commands save a *Maple* worksheet to an existing or new file. These commands do *not* by default save the internal state. In other words, when *Maple V* is started with a previously saved worksheet, the mathematical engine knows nothing of the mathematics of that worksheet until the appropriate input regions have been re-executed or the associated variables redefined. In any case, the worksheet as saved may not accurately represent the internal state at the time it was saved.

To save the internal state at any time, you must select the **Save Kernel State** option from the **Options** menu and then **Save** or **Save As....** Saving the internal state saves *Maple*'s internally defined variables and user's variables and allows you to reload the session and continue working with all your variables and procedures defined.

Note: A worksheet that includes *Maple*'s internal state always requires more disk space than a normal worksheet. This is because two files must be saved: a file with a **.ms** extension (the worksheet representation) and a file with a **.m** extension (the internal state). Therefore, we recommend that you use the **Save Kernel State** option only when needed.

If you do not use the **Save Kernel State** option when saving a worksheet, you can recover the previous state of *Maple V* by loading in the worksheet, removing all output, and then re-executing each input line or choosing **Execute Worksheet** from the **Format** menu.

Users of earlier releases of *Maple V* for Windows should note that *Maple V Release 3* for Windows will read all your worksheets from previous releases. For worksheets from previous releases that were saved with the **Save Kernel State** option, *Maple V Release 3* for Windows will be able to read these worksheets from the **.ms** file, but it will be unable to read the saved kernel state from the **.m** file. However *Release 2* **.m** files can be converted into a form compatible with *Release 3* by using the **m2src** application described on page 61.

Types of Maple V Windows

There are five different types of windows that can be opened during a *Maple V* session. All the windows created by *Maple V* can be moved around the screen in the standard Windows fashion.

Worksheet Window

All *Maple* commands to be calculated are entered in the worksheet window, and all mathematical results are also displayed in this window. Only *one* worksheet window can be open during any *Maple* session. The worksheet window controls the session. For more information, see the section *Working with Worksheets* on page 11.

Help Window

A help window displays a *Maple* help page for a specific command, package, or structure. Multiple help windows can be open concurrently. For more information, see the section *Maple V's Help Facility* on page 28.

Two-dimensional Plot Window

A two-dimensional plot window displays a plot generated by the **plot** command. Multiple two-dimensional plot windows can be open concurrently. For more information, see the section *Manipulating Plots* on page 17.

Three-dimensional Plot Window

A three-dimensional plot window displays a plot generated by the **plot3d** command. Multiple three-dimensional plot windows can be open concurrently. The plots in these windows can be manipulated through the use of the 3D plot menus, buttons, and the mouse. For more information, see the section *Manipulating Plots* on page 17.

Animation Window

An animation window is a variation of a two-dimensional or three-dimensional plot window in which an animation is run. Animations are generated by the **animate** and **animate3d** commands and by the **display** and **display3d** commands with the option **insequence=true**. Multiple animation windows can be open concurrently. For more information, see the section *Animation* on page 22.

Working with Worksheets

The user interface for *Maple V* for Windows is based on the concept of *worksheets*. The worksheet shows the input, output, text, and plots created in a session. When *Maple V* for Windows is started, a new blank worksheet is created. To load in an existing worksheet, use the **Open** menu item in the **File** menu. The information in a worksheet is fully editable.

The four types of regions that can be included in a worksheet are input, output, text, and graphics. You cannot physically manipulate output or graphics regions; however, you can edit text and input regions. Furthermore, you can convert text regions to input regions and vice versa through the use of the **Text/Input Region** item in the **Format**

Worksheet Regions

menu or by pressing **F5**. A sample worksheet window is displayed below.

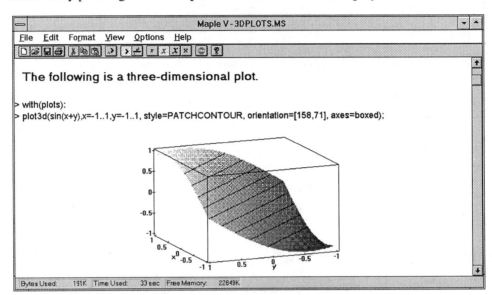

At the top of the worksheet are the **Menu Bar** and the **Tool Bar**. The function of each button in the **Tool Bar** is shown at the bottom of the worksheet by placing the cursor over the button and holding down the left button on the mouse. A summary of the functions of all buttons in the worksheet and in plot windows is given in the help page under **Interface Help**, **What's New in the Windows Interface**, and **Tool Bars**.

At the bottom of the worksheet window is the **Status Bar**. Normally, this bar displays information about resources used by *Maple*, such as **Bytes Used**, **Time Used**, and **Free Memory**. When a menu item or **Tool Bar** item is chosen, the **Status Bar** displays how the chosen option operates.

To see a sample *Maple V* for Windows worksheet, select **Open** in the **File** menu and open the worksheet **QUIKTOUR.MS** located in the directory where you installed *Maple V Release 3*.

Worksheet Regions

All the information displayed in a worksheet is arranged into regions. There are four distinct types of regions: input, text, output, and graphics. A region can be of any length. The various types of regions are discussed in more depth in the following sections.

Input Regions

Input regions are used for entering commands to be executed by Maple's mathematical engine. A *Maple V* session starts in an input region. In an input region, the > character marks the left side of the current line. In such a region, typing a command followed by a semicolon and pressing **Return** or **Enter** will send the command to be calculated. You can also double-click anywhere in the input region with the left mouse button to send an input region to be calculated.

Return versus Shift + Return

The **Return** and **Shift + Return** key sequences have different functions in *Maple V* for Windows.

In an input region, the **Return** key always tries to send input to the mathematical engine to be calculated.

The **Shift + Return** key sequence inserts a carriage return and moves the input cursor to the lower line. It does not send input to be calculated. Pressing **Shift + Return** allows you to break lines or type additional lines of input before sending the command(s) to the mathematical engine. These multiple lines are part of the same input region.

Input Prompts

A prompt character > appears at the left margin of the worksheet's input line. Input prompts can be toggled on and off in the **View** menu, by pressing the button on the worksheet **Tool Bar**, or by pressing the **F10** key.

Note: *Maple* recognizes only input prompts that it has created itself. A > that you just type in directly will *not* be recognized as an input prompt.

Text Regions

Text regions contain textual information that is not meant to be calculated in your worksheet.

Text regions can be placed anywhere in a worksheet. The most common place for text is directly before a *Maple* input region. To create a text region, place your cursor on a blank line and press **F5**. To terminate a text region and transfer into an input region, place your cursor on a blank line and press **F5**. This toggling can also be done with the **Text/Input Region** item in the **Format** menu.

Output Regions

Output regions display results of *Maple* commands that have been entered through an input region. Some *Maple* commands generate no screen output, whereas others create several pages. Output is placed in an output region associated with the input command. The position of the output region with respect to subsequent regions can, within limits, be controlled by the user using *region groups* described below.

Text and input regions can be converted through the **Text/Input Region** item in the **Format** menu or by pressing **F5**. Output regions, ordinarily displayed using graphics-based mathematical typesetting, cannot be converted into either input or text regions. However, **interface** commands can be used to select a text output display that allows some copying of output text into input and text regions.

Graphics Regions

Plots from two-dimensional plot windows or three-dimensional plot windows can be pasted into a worksheet. A *Maple* plot pasted into a worksheet can no longer be manipulated by *Maple*. Any amendments to the plot must be regenerated in the plot window.

Pictures and graphics from other Windows applications can also be pasted into the worksheet via **Cut/Copy/Paste**. Also, plots created by Maple V can be copied from plot windows and pasted into other Windows applications.

Region Groups

In addition to the individual regions just described, *Maple V* for Windows worksheets also have *region groups*. A region group may consists of an input region, its related output region, and any adjacent text region. To see where the region groups are, turn on **Separators** in the **View** menu, press the button in the **Tool Bar**, or press **F9**.

Note: If you place the cursor anywhere within the input region of a region group and press **Enter** or double-click the left mouse button, you will automatically re-execute all the input lines in that region group.

The **Split Group** item in the **Format** menu is used to split up existing region groups, and the **Join Group** item is used to join existing region groups. To insert a new command block either above or below the current one, select **Insert New Region**.

Editing Worksheets

At any time during a *Maple V* session, input and text regions can be edited. However, editing existing input can create a worksheet in which the mathematical output does not make sense in the sequence in which it appears. The following two figures illustrate this point.

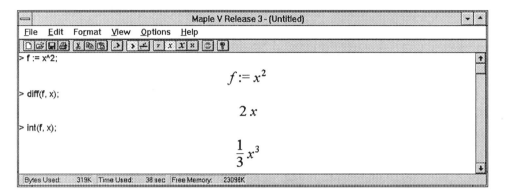

After creating a worksheet with a sequence of *Maple* calculations using a function **f**, the user changes the first input region from **f:=x^2;** to **f:=2*x;** and re-executes only this one new input region. The worksheet is then saved. The next time this worksheet is opened, the subsequent calculations involving **f** seem to be incorrectly calculated because of the change.

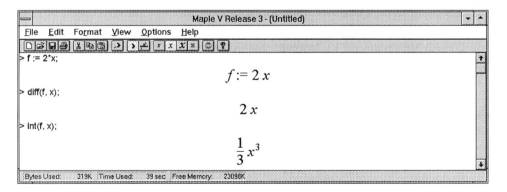

It is also possible to edit a worksheet and alter input without *any* re-execution. This also creates misleading worksheets. Hence, common sense and caution should be used when editing the input regions of worksheets. If the results displayed in a worksheet are suspect, the entire worksheet should be re-executed or recalculated.

Recalculating a Worksheet

The worksheet-based user interface of *Maple V* for Windows allows you to execute an entire *Maple V* worksheet after making mathematical changes to any input statements.

To recalculate an entire worksheet, first ensure that **Continuous Mode** is switched on in the **Options** menu. Then select **Execute Worksheet** in the **Format** menu. All input regions are executed in the sequence in which they appear in the worksheet. If the input in the session is entered or amended out of sequence, recalculating the entire worksheet may lead to misleading output.

By turning on **Replace Mode** in the **Options** menu, you can have newly calculated output replace the previously calculated output. Plots defined in the worksheet are recalculated, but the new plots do not automatically replace previous plots pasted in the worksheet. For more information, see the section *Manipulating Plots* on page 17.

If **Replace Mode** is not turned on, then all newly calculated output is inserted before the previously calculated output. This is useful when you change an input line slightly and want to compare the new results with the previous results.

The **Continuous Mode** item in the **Options** menu allows you to move from one region group to the next. If **Insert Mode** is set on, then pressing **Enter** at an input region inserts a new region group immediately following the current region group. The final option available is **Insert At End Mode**. In this mode, pressing **Enter** at an input region inserts a new input region at the end of the worksheet and moves the cursor to that position.

Converting Maple V Worksheets

Between Release 2 and Release 3

A *Maple* worksheet that was created using *Maple V Release 2* for DOS/Windows is readable by *Maple V Release 3* for DOS/Windows. Launch *Maple V Release 3*, select **Open** from the **File** menu, and indicate the appropriate **.ms** file. Your *Maple V Release 2* worksheet will be automatically converted to *Maple V Release 3*.

The worksheet is not completely converted, however, until you save it as a *Maple V*

Release 3 worksheet. Use caution, and back up or rename such files: once a worksheet has been converted, *Maple V Release 2* will no longer be able to read the new *Maple V Release 3* worksheet.

For more information, see the section *Saving Maple's Internal State* on page 9.

Among Release 3 Platforms

If you run *Maple V Release 3* on a different platform and want to convert those worksheets to *Maple V Release 3* for DOS/Windows, one alternative is to follow the same procedure used for converting *Release 2* worksheets to *Release 3*. That is, opening another platform's standard format, *Release 3* worksheet in *Maple V Release 3* for DOS/Windows converts it automatically.

Note: Files saved using the **.ms** extension in DOS/Windows are "standard format" files. See the section *Saving the Session* on page 23. The procedure differs by platform, however, so refer to the other platform's documentation for information on how to save standard format files there. (In *Getting Started* for the Macintosh, see **Save a Copy In...** in the section *Saving the Session* on page 29.)

A second alternative is to transfer the worksheet as a plain ASCII text file between both platforms, using files saved with the **.txt** extension. See the section *Saving the Session* on page 23. (In *Getting Started* for the Macintosh, see the section *Importing and Exporting* on page 30.)

The second method is recommended if your worksheet contains large amounts of text. The first method is recommended if your worksheet contains large amounts of multiline input, such as procedures.

Manipulating Plots

Maple V includes commands for creating both two-dimensional and three-dimensional plots. The principal *Maple* commands for plotting are **plot** and **plot3d**. Several other commands provide special classes of plots—for example, **tubeplot, sphereplot, animate, animate3d**, and so on. For more information on these plotting commands, see the *Maple V Quick Reference* or the associated help pages.

Plots are created, named, and manipulated much like any other algebraic object in

Manipulating Plots

Maple, but they are displayed in separate windows. These windows can be moved around the screen, resized, and closed in standard Windows fashion. Resizing a plot window redraws the plot to fill the new window size. Multiple plot windows can be open concurrently or closed at any time. If a plot window has been closed, the plot can be redisplayed either by referring to the plot object by name or by re-entering the appropriate plot command from the worksheet.

Plots can also be pasted into a worksheet. With the plot window as the active window, copy the plot using **Copy** in the **Edit** menu or **Ctrl + C**. Then, with a worksheet window as the active window and the worksheet's insertion point located where the graphic is to be placed, select **Paste** in the **Edit** menu or press **Ctrl + V**. You can also paste *Maple V* plots in other Windows applications. Much of this work can also be performed using the plot window **Tool Bar**.

A plot pasted into a worksheet can no longer be manipulated by *Maple V*. If you wish to change the appearance of the pasted plot—for example, its lighting or rendering scheme—make the appropriate changes in the original plot window and recopy the plot into the worksheet.

Two-dimensional Plots

Two-dimensional plots are created principally with the **plot** command. Changes to the appearance of a two-dimensional plot are controlled through parameters of the **plot** command or through the menu items available in the two-dimensional plot window. A sample two-dimensional plot window is shown here.

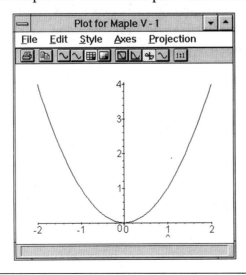

- The **File** menu allows you to **Exit** or quit the plot window or **Print** the plot. The **Print** option allows you to print to a printer or to save the plot in a file. The **Printer Setup...** option allows you to control how your printer displays information.

- The **Edit** menu allows you to **Copy** the plot to the Clipboard so that it can be pasted into a worksheet or any other Windows application.

- The **Style** menu allows you to view the plot in either **Line, Point, Patch w/o Grid**, or **Patch** mode. The default is **Patch** mode. In most cases, **Patch** mode and **Line** mode are the same; however, certain plot commands, such as **densityplot**, have a different representation in **Patch** mode versus **Line** mode.

  The **Style** menu also has submenus for setting **Symbol, Line Style**, and **Line Width**. Available symbols are **Cross, Diamond, Point, Circle, Box**, and **Default** (a small **Cross**). Available line styles are **Solid, Dash, Dot, DashDot, DashDotDot**, and **Default** (**Solid**). Finally, available line widths are **Thin, Medium, Thick**, and **Default** (**Thin**). **Note:** Some combinations of **Symbol, Line Style**, and **Line Width** are not simultaneously available.

  Also, the **Tool Bar** and **Status Bar** can be toggled on and off.

- The **Axes** menu allows you to put **Boxed, Framed**, or **Normal** axes on the graph, or to have no axes displayed. The default is to have **Normal** axes displayed.

- The **Projection** menu allows you to view the plot with **Constrained** or **Unconstrained** scaling. **Constrained** scaling causes equal scaling of both axes. **Unconstrained** scaling adjusts the axes' scale to fit the plot window. The default is to display the plot **Unconstrained**.

The **Tool Bar** at the top of the plot window provides quick access to some of the above menu commands. There are buttons for printing the plot, copying the plot to the Clipboard, changing the plot style, changing the axes' style, and toggling between projections.

The **Status Bar** at the bottom of the plot window typically displays the x and y location of the cursor when the mouse is clicked within the plot window. When a menu item or **Tool Bar** item is selected, the **Status Bar** is used to provide a description of the item. A summary of all button functions in the plot windows is given in the help page under **Interface Help, What's New in the Windows Interface, Tool Bars**.

Three-dimensional Plots

Three-dimensional plots are created principally with the **plot3d** command. Changes to the appearance of a three-dimensional plot are controlled through parameters to the **plot3d** command or through the menu items available in the three-dimensional plot window.

Note: If you re-create a three-dimensional plot by re-executing the input region that originally created the plot, the plot is displayed without any previous changes that were made via the three-dimensional plot menus. For each new plot that is created, the default values of the options in the three-dimensional plot menus are applied. A sample of a three-dimensional plot window is shown here.

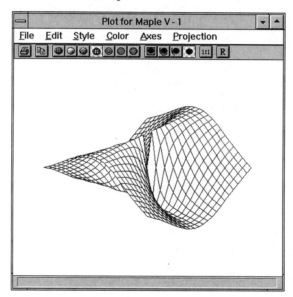

Rotating a Plot

To rotate a three-dimensional plot, click the mouse on any part of the plot itself and a rotation box appears. Dragging the mouse around the screen while the left mouse button is still pressed rotates the box. You can also rotate the box by using the arrow keys on the keyboard, which actually give you more precise control over the rotation of the graph.

To redraw a three-dimensional plot, either double-click the left mouse button, click the right mouse button, press the **Redraw** button on the **Tool Bar**, or press **Enter** on the keyboard.

The 3D Plot Menus

- The **File** menu allows you to **Exit** or quit the plot window or **Print** the plot. The **Print** option allows you to print to a printer or save the plot to a file.

- The **Edit** menu allows you to **Copy** the plot to the Clipboard so that it can be pasted into the worksheet or in any other Windows application. The **Edit** menu also allows you to **Redraw** the three-dimensional plot.

- The **Style** menu allows you to view the plot in many different styles. The default is to display the plot in **Patch** mode with a grid on the surface; however, you can also display the graph in patch mode without a grid (**Patch w/o grid**) or patch mode with contours (**Patch and contour**).

  The **Style** menu also has submenus for setting **Symbol**, **Line Width**, **Line Style**, and **Grid Style**. Available symbols are **Cross**, **Diamond**, **Point**, **Circle**, **Box**, and **Default** (a small **Cross**). Available line styles are **Solid**, **Dash**, **Dot**, **DashDot**, **DashDotDot**, and **Default** (**Solid**). Available line widths are **Thin**, **Medium**, **Thick**, and **Default** (**Thin**). Finally, available grid styles are **Full Grid** (triangular) and **Half Grid** (rectangular).

  Note: Symbol options do not work with **Line** mode, and **Line Style**, **Line Width**, and **Grid Style** options do not work with **Point** mode.

  Also, the **Tool Bar** and **Status Bar** can be toggled on and off.

- The **Color** menu allows you to change the color scheme of the graph and add lighting to the graph. **Light Schemes 1, 2, 3,** and **4** are predefined lighting schemes set for your convenience. If you specify your own lighting scheme on the command line that created the three-dimensional plot, then the **User Lighting** option in the Color menu becomes available so that you can toggle between **No Lighting**, **User Lighting**, and **Light Scheme 1, 2, 3,** or **4**.

  One of the items available in the **Color** menu is **Dither**. Dithering blends the colors in your plot at the region boundaries. The dithered image produced is more portable and will be more accurately reproduced when pasted into a worksheet. **Dither** is available only if you have a display adapter driver that supports 256 colors or more.

- The **Axes** menu allows you to put **Boxed**, **Framed**, or **Normal** axes on the graph, or to have no axes displayed. The default is to have no axes displayed.

- The **Projection** menu allows you to view the plot with **Constrained** or **Unconstrained** scaling. The default is to display the plot **Unconstrained**.

Manipulating Plots

The **Projection** menu also allows you to view the plot from **Near**, **Medium**, **Far**, or **No Perspective**. The default is to display the graph with **No Perspective**.

The **Tool Bar** at the top of the plot window provides quick access to some of the above menu commands. There are buttons for printing the plot, copying the plot to the Clipboard, changing the plot style, changing the axes' style, toggling between projections, and redrawing the plot. A summary of all button functions in the plot windows is given in the help page under **Interface Help**, **What's New in the Windows Interface**, **Tool Bars**.

The **Status Bar** at the bottom of the plot window typically displays the current rotation values for your three-dimensional plot. When a menu item or **Tool Bar** item is chosen, this location is used to provide a description of the item.

Animation

An animation window is a variation of a two-dimensional or three-dimensional plot window. When you create an animation using either the **animate** or **animate3d** command, a window similar to the following example is displayed:

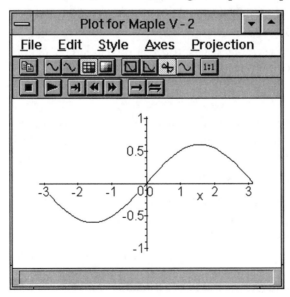

Notice that all the normal two-dimensional or three-dimensional menu items are available in an animation window. The animation window also displays a set of VCR-style buttons used to either scan or play the animation. In play-once mode, the animation plays through all frames once and then stops. In loop mode, the animation replays until you stop it. A summary of all animation window VCR-button functions is in the help page under **Interface Help**, **Plotting in Maple**, **Animation**.

Maple V Menus

This section looks at the menus that are available for a *Maple V* worksheet. The commands that appear in each menu are grouped so that, as much as possible, commands that do similar tasks appear in the same menu. At times, some of the commands listed in the menus are dimmed, when their actions are not appropriate.

File Menu

The **File** menu contains commands for manipulating *Maple* files in the Windows system. Files can be opened, saved, or printed by using the commands in the **File** menu. Also, *Maple V* can be terminated by using the **Exit** command (**Alt + F4**).

Selecting **New** removes the current worksheet and restarts *Maple V* with a new worksheet, having executed a Maple **restart** command. If the current worksheet has not been saved since editing, *Maple V* first prompts you to save the worksheet and then executes the **New** option.

Selecting **Open...** opens a dialog box that asks for the location of the worksheet that is to be opened. When a worksheet is selected, **Open...** functions like **New** except that once the current worksheet is removed, the newly selected worksheet is loaded and displayed. If the newly selected worksheet was previously saved with the **Save Kernel State** option, then *Maple V* reinitializes its internal state to match the state when that worksheet was saved.

Saving Options

Selecting **Save Settings** saves any options that are currently set in the **Options** menu so that when you restart *Maple V* for Windows, your options will automatically be set.

Saving the Session

The commands **Save** and **Save As...** are standard Windows commands that allow you to save the contents of a worksheet.

Maple V Menus

Save saves a copy of the current worksheet by overwriting the previously saved copy of that worksheet.

Save As... opens a dialog box that asks for the location and the new name of the file to be saved. Also, you can specify which file type you want to use:

- **.ms** format (*Maple V*'s internal worksheet format)

- **.tex** format (for exporting to a LaTeX document),

- **.txt** format (for saving as plain ASCII)

- Any other extension you wish to provide that will save as plain ASCII.

The *Maple V* window then displays the newly saved file.

Note: When saving a worksheet, you must use the **.ms** extension to tell *Maple V* that you are saving the current worksheet as a *Maple V* session. If you chose the **Save Kernel State** option prior to doing a **Save** or **Save As...**, then a second file with a **.m** extension is automatically created to contain the saved internal state. If you do not specify the **.ms** extension, then the worksheet is saved as a text file.

Printing Files

Worksheets, plots, and help pages can be printed on a printer connected to your computer. Selecting **Print...** brings up the Windows printing dialog box and allows you to print your file. The **Print** option allows you to print to a printer or to a file.

Selecting **Printer Setup...** opens a standard Windows dialog box to customize the setup of your printed pages.

Selecting the **Page Margins...** item in the **File** menu brings up a dialog box that allows you to set page margins for your printouts.

Edit Menu

The commands **Cut**, **Copy**, **Paste**, and **Delete** are standard Windows commands that allow you to manipulate objects in the *Maple V* session. They work alongside the current selection in a manner similar to that of many other Windows applications.

- **Cut (Ctrl + X)** places a copy of the current selection on the Clipboard and deletes it from the document.

- **Copy (Ctrl + C)** places a copy of the current selection on the Clipboard without deleting the selection. If a plot window is the active window, it is not necessary to select the plot before using the **Copy** command.

- **Paste (Ctrl + V)** replaces the current selection with the contents of the Clipboard. If you want to **Paste** a plot directly following the command that created it, select any character in the command and select **Paste**. The plot will be pasted after the command.

- **Delete (Delete)** deletes the current selection without affecting the Clipboard.

- **Copy to...** allows you to place a copy of the current selection in another file in **.ms**, **.tex**, or **.txt** file format.

Format Menu

The **Format** menu contains commands for formatting worksheets and individual regions within a worksheet. Other available commands allow you to toggle between input and text regions, select fonts, and execute the entire worksheet.

- **Text Region/Input Region (F5)** takes the current input region and converts it into a text region, or vice versa.

- **Split Group(F3)** splits the current region group into two region groups.

- **Join Group(F4)** joins the current region group with the group above it.

- **Insert Page Break** inserts a permanent page break at the current cursor location. This page break remains in place until removed.

- **Insert New Region** inserts a new region either **Above (Ctrl + O)** or **Below (Ctrl + I)** the current region.

- **Remove All** removes all the input, output, text, or graphics in a worksheet. **Note:** You cannot remove the type of region in which the cursor currently lies.

- **Fonts** allows you to change the font, color, and size of all input, output, and text regions. When you save a document, it is saved with the current font settings.

Maple V Menus

- **Math Style** allows you to change the size of the standard math notation output that is generated by *Maple V Release 3*. The size choices are **Large**, **Medium**, and **Small**. If you change the color and size of the output font, only the *color* of the standard math notation is changed. You must use the **Math Style** option to change the *size* of the standard math notation. The only other option available for **Math Style** is **Character**. Choosing **Character** style will change all subsequent output generation to the pretty-printed style. To change the color and size of this **Character** style output, you must use the **Fonts** option.

- **Execute Worksheet** allows you to recalculate all the regions in a worksheet if **Continuous Mode** is checked in the **Options** menu. For more information, see *Recalculating a Worksheet* on page 16.

View Menu

The **View** menu contains commands for controlling the appearance of the worksheet and the individual regions within a worksheet.

- Turning on **Separator Lines (F9)** displays the horizontal lines that separate region groups in a worksheet. Turning off **Separator Lines** removes any such lines from a worksheet.

- Turning on **Input Prompts (F10)** displays the prompt character > at the start of all input lines.

- Turning on **Status Bar (F2)** displays on the screen a **Status Bar** that indicates the memory and CPU usage of *Maple V*. Turning off the **Status Bar** hides it.

    Note: The memory indicator of the **Status Bar** is updated only after each *Maple* garbage collection. For more information on garbage collection (gc), consult the relevant sections in *Maple V Quick Reference* (by Blachman and Mossinghoff).

- Turning on **Tool Bar** displays the **Tool Bar** at the top of the worksheet window. Turning off **Tool Bar** removes it.

- Turning on **Plot Tool Bars** displays the **Tool Bars** at the top of both two-dimensional and three-dimensional plot windows. Turning off **Plot Tool Bars** removes them. These changes affect only subsequently opened plot windows.

Options Menu

The **Options** menu contains commands for specifying how a worksheet operates. **Note:** To save your options so that they will be automatically set up when *Maple V* is restarted, use the **Save Settings** item in the **File** menu.

- Turnning on **Save Kernel State** will save the internal state of *Maple V* when the worksheet is saved. Turning off **Save Kernel State** means that the worksheet contents alone are saved when **Save** or **Save As..** is selected.

- **Confirmation Checks** allows you turn off all "are you sure..." prompts. For example, if you decide to **Exit** the current session, or if you decide to **Remove All** input, then by default you are prompted to save any current changes or verify that you really do want to remove all input. Use caution when you turn off **Confirmation Checks**.

- **Fast Graphics Redraw** causes *Maple V* to attempt to keep in memory all three-dimensional plots so that they can be redrawn quickly. This is useful if you intend to display several three-dimensional plot windows and then quickly change the active plot window. **Note:** By turning **Fast Graphics Redraw** on, you begin to use up more memory.

- **Automatic Save Settings** causes *Maple V* to save any options that are currently set at exit time, so you do not have to first select **Save Settings** from the **File** menu.

- Turning on **Replace Mode** forces all new output to replace any previous output in a region group. This is the default. If **Replace Mode** is turned off, then the output regions are retained for comparison purposes.

- **Continuous Mode** moves execution from one region group to the next when recalculating inputs, whereas **Insert Mode** inserts new regions between the current and the next region, and places the cursor in the new region. **Insert at End Mode** inserts a new region and places the cursor at the end of the worksheet.

Help Menu

The **Help** menu has five menu items:

- The first item, **Browser...** (**F1**), displays a help file browser dialog box. For more information see the section *Maple V's Help Facility* on page 28.

Maple V's Help Facility

- Another way of selecting a help topic is through **Keyword Search...** (**Shift + F2**). For more information see the section *Maple V's Help Facility* on page 28.

- **Interface Help...** (**Shift + F1**) displays Windows Help on the *Maple V* for Windows interface. These help pages supplement this guide.

- **Help on "..."** (**Ctrl + F1**) displays the *Maple* help page for the command currently highlighted in the *Maple V* session and may be referred to as *context-sensitive help*.

- Selecting **About Maple V...** brings up a dialog box indicating the current version of *Maple V* for Windows and the copyright notice. Your name and serial number are also displayed in this dialog box.

Maple V's Help Facility

Maple V has an extensive facility for helping you understand its language, procedures, and syntax. For more information on the structure of help pages, see the *Maple V Quick Reference* or view the help pages themselves. There are four ways to access a help page during a session, as outlined in the following paragraphs.

Selecting **Browser...** (**F1**) in the **Help** menu opens a help page browser dialog box. *Maple*'s help pages are organized in a logical hierarchy with, at most, five levels of reference. A topic with three dots (...) following it has related subtopics beneath it. The leftmost listing is the highest level of reference and contains several general topics—for example, **Mathematics**, **Graphics**. Each list to the right is one level of reference lower. A picture of the help browser is shown on the following page.

Selecting (with a single click) a topic with three dots opens a list of the related subtopics. Also, selecting any lowest-level topic updates the *synopsis* region, located at the bottom of the browser, with a brief description of the help page for that topic. To open a help page from the dialog box, either double-click the left mouse button directly on a topic, or select a topic and click the **Help** button. The browser is particularly useful when you are not sure of the exact name of the command you require. Multiple *Maple V* help pages can be open in this way, and each is displayed in a separate window.

The second way to access *Maple* help pages is with the **Keyword Search...** (**Shift + F2**) facility. In the **Search for String** field, type a topic or character string you want to search for and click the **Search** button. The synopsis fields from *all* of *Maple V*'s help pages are searched for the appropriate characters, and a list of applicable help pages is displayed in the **Help Topics** area. Double-click the topic you want, or highlight it and click the **Help Page** button, to select that individual help page.

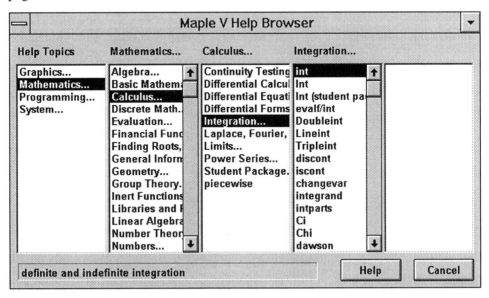

The third way to access a *Maple V* help page is to place the cursor at any point within the term for which you want help, and then select **Help on ...** (**Ctrl + F1**) from the **Help** menu. The appropriate help page is displayed. This option is a shortcut for the **help()** and **?<topic>** syntax.

The fourth way to access the help facility is by issuing a help command with the **?<topic>** syntax directly from an input region of the worksheet. For example, try **?simplify**.

Using Maple V Release 3 for DOS

Objectives of This Chapter

This chapter describes the user interface features of the DOS version of *Maple V Release 3*. It is assumed that *Maple V* is correctly installed and running on your computer. To install *Maple V* on your system, refer to the chapter *Installing Maple V Release 3* on page 3.

The DOS version of *Maple V Release 3* offers a user interface packed with features designed to increase productivity:

- A graphics display driver for two- and three-dimensional plots is included, with menus that allow plotting options to be changed on the fly. By using the menus you can replot with different style options and rotation angles until you get the plot you want.

- *Maple V* can be driven from a menu system. Common operations such as loading library packages and simplifying expressions can be done from the menus, which are fully user-definable. This facility allows menu items to be added, removed, and modified.

- A full text editor for creating *Maple* expressions and files is included. **MAPLEDIT** can be called from within *Maple V*, allowing interactive development of procedures.

- A log of the current session is automatically kept every time *Maple V* is run. By pressing a single key, you can enter the session browser and review an up-to-date copy of the session log. This copy of the session log can be edited using MAPLEDIT. Before exiting the browser, you can select a block of lines to be read back into the session for re-execution.

- An input/output capture mode is provided. Using this facility, you can tell *Maple V* when to start and stop saving a copy of the session log input and/or output to a file.

Starting Maple V

- A **Save/Load Session** facility preserves the exact state of your session between uses of the program. Every aspect of the session—including the session log, defined values and procedures, and the command list—is preserved.

- A command line editor lets you change the line you are typing and recall the last 100 typed lines for editing and re-execution. A search facility finds a specific line when you type in a prefix of that line.

- An interactive help topic browser is included, allowing you to move through a logical hierarchy of *Maple*'s numerous help pages. One-line synopses of the topics allow you to find the appropriate information easily. An additional searching mechanism is provided to search for all synopses that contain a certain string.

These features have been designed to provide flexibility and ease of use. For example, a more advanced user can set up a menu that makes it easy for beginners to get started on a particular topic. The browser is used for displaying help files and reviewing a copy of the session log. The same editor is used for editing expressions, *Maple V* files, and copies of the session log.

Starting Maple V

To start *Maple V*, type **MAPLE** at the DOS prompt and press **Return**.

You can now begin typing commands. A good, quick way to see *Maple V* in action is to load help pages into the browser and select command line examples to be read into the session for execution. The section *Maple V's Help Facility* on page 54 explains how to do this.

When you first start a session, the screen appears as shown in the following diagram.

Exiting Maple V

To exit *Maple V*, press the **F3** key. A window appears, asking you to confirm that you wish to exit. Press **Y** to confirm.

Interrupting a Maple Computation

Use **Ctrl + C** or **Ctrl + Break** to interrupt *Maple V* during a long computation. The computation is interrupted and the input prompt appears.

Maple Initialization Files

If a file named **MAPLE.INI** exists in the *Maple* **LIB** directory, it is processed by *Maple V* whenever the application is started. This file contains commands to be executed before the first prompt appears. After this, the **MAPLE.INI** file in the current directory, if any, is processed. Use these files to set up *Maple V* the way you like. For example, your **\MAPLEV3\LIB\MAPLE.INI** file as supplied contains the following:

Maple Initialization Files

```
# The following lines are read by Maple's DOS interface BEFORE the
# Maple kernel is started.  These lines, if they are to be processed,
# must appear before any non-comment lines in this file.
#
# Maximum number of saved lines in session:
# SESSMAX=1000
#
# Maple colors:
## MAPLECOL=0E0007F
#
# Default hardcopy plot device and port (when printing from plot menu):
# MAPLEPRT=laserjet
# MAPLEPRF=LPT1
#
# The following should only be used if you want to run Maple in graphics
# mode instead of text mode.  If so, change the "##" to "#" in the lines
# you want to activate.
#
# The video mode to use (18 is 640x480 16 color):
## VIDEOMD=18
#
# The foreground and background colors, expressed as three 2-digit
# hexadecimal numbers from 00 to 3F, representing blue, green, and red
# intensities:
## FOREGR=0x000000
## BACKGR=0x2A3939

# Everything from here on is Maple (instead of interface) initialization.

# If this file has been read already, do not read it again.  This
# can only happen if you run Maple from inside its LIB directory.
if `sys`init`file`read = 1
then `sys`init`file`read := ``sys`init`file`read`
else `sys`init`file`read := 1:

# A value of 5000 causes the status line to be updated reasonably
# often.  Smaller values will cause more frequent updates.  Larger
# values will cause less frequent updates, but may improve performance.
words(5000):

# These are needed so that fed, fred, and xed are available.
readlib(fed):
readlib(fred):
readlib(xed):

# Fill up `, ``, and ``` with zero so we have a known state.  This is
# needed for the Save Session menu item to work properly.
0:0:0:

fi:
```

This file contains comments that explain the individual commands. Many of the commands in the **MAPLE.INI** file make *Maple V* run efficiently on your system. If you want to add to the **MAPLE.INI** file, it is recommended that you do not delete any of the current contents but rather append commands to the file.

The first few lines of the **MAPLE.INI** file are used to set the *Maple* environment variables **SESSMAX**, **MAPLECOL**, **MAPLEPRT**, **MAPLEPRF**, **VIDEOMD**, **FOREGR**, and **BACKGR**. In general, any environment variable that appears in **MAPLE.INI** in the form

```
# ENVIRONMENTVAR=VALUE
```

overrides any values set from the DOS prompt or the **AUTOEXEC.BAT** file. The three environment variables **VIDEOMD**, **FOREGR**, and **BACKGR** appear with two # signs in front of them and are used only if you remove one of the # signs. For more information on *Maple* environment variables and their uses, please see the section *Maple Environment Variables* on page 36.

You can create a **MAPLE.INI** file using any text editor or a word processor in nondocument mode. *Maple V* includes such a text editor. To bring up this editor to edit a file, issue the following command at the DOS prompt:

MAPLEDIT *filename*

Alternatively, you can invoke the same text editor from within *Maple* with the command

fed(`** *filename* **`);

where ` is the back-quote character.

See the section *Editing in Maple V* on page 42 for details on how to use **MAPLEDIT**.

Multiple Initialization Files

The *Maple V Release 3* installation process automatically creates one or two different initialization files for you:

C:\WINDOWS\maple3.ini, a Windows-specific initilization file, and

C:\MAPLEV3\LIB\maple.ini, described above.

Getting Started

These set various defaults, and should be left in place and intact, though they may be expanded. Your own customizations, such as resetting the global variables Digits or Order, are best kept in a file in your specific working directory for Maple, such as:

C:\MAPLEV3\BIN\maple.ini

if that specifies the working directory shown in Properties for *Maple V Release 3*, or the directory from which you issued the DOS **maple** command.

Maple Environment Variables

MAPLECOL

By default, on a color screen *Maple V* uses a red/yellow/black color scheme. If you do not like this color scheme, design your own by setting a DOS environment variable called **MAPLECOL**. This variable should be set as a string of seven hexadecimal digits. A hexadecimal digit is a numeral from 0 to 9 or a letter from A to F. These digits specify the following attributes:

1st digit	Session background color (0–7)
2nd digit	Session text color (0–F)
3rd digit	Ignored in *Maple V Release 3*
4th digit	Ignored in *Maple V Release 3*
5th digit	Ignored in *Maple V Release 3*
6th digit	Reverse video background color (0–7)
7th digit	Underline color (0–F)

To remain compatible with environment variables that might have been set in previous versions of *Maple*, the **MAPLECOL** variable is still seven digits long, even though the middle three digits are ignored.

The reverse video and underline colors are used for, among other things, menu selection and menu shortcut keystrokes, respectively.

The hexadecimal digits represent the following colors. Actual colors displayed depend on your monitor.

0	Black		8	Dark Gray
1	Blue (Underline on monochrome display)		9	Bright Blue
2	Green		A	Bright Green

3	Cyan (Turquoise)	B	Bright Cyan
4	Red	C	Bright Red
5	Magenta (Hot Pink)	D	Bright Magenta
6	Brown	E	Yellow
7	Light Gray	F	White

For example, the following statement in your **AUTOEXEC.BAT** file produces colors similar to the default colors:

SET MAPLECOL=070004E

If you put such a statement in your **AUTOEXEC.BAT** file, it does not take effect until you reboot your computer. You can also type such a statement at the DOS command line before running *Maple V* to experiment with the colors without repeated rebooting.

Note: The above information applies specifically to color displays. If you install *Maple V* on a computer with a monochrome display, **MAPLECOL** will automatically be set to a value that is legible on such a display.

VIDEOMD

A second environment variable that is used by *Maple V Release 3* for DOS is **VIDEOMD**. By default, *Maple V* runs in text mode and switches to the best available "standard" graphics mode when doing plots. The best graphics mode you can get this way is 640×480 with 256 colors. If you specify the environment variable **VIDEOMD=n**, for some *n*, *Maple V* runs in a specified graphics mode. This does not work with all display adapters. The supported values for *n* are:

11	720×350 Hercules monochrome
13	320×200 16 colors
14	640×200 16 colors
15	640×350 monochrome
16	640×350 4 or 16 colors
17	640×480 monochrome
18	640×480 16 colors
19	320×200 256 colors
256	640×400 256 colors
257	640×480 256 colors
258	800×600 16 colors
259	800×600 256 colors
260	1024×768 16 colors

261   1024×768 256 colors

Note: In graphics mode, the **MAPLECOL** variable is ignored and the **FOREGR** and **BACKGR** variables (see below) are used instead.

FOREGR=XXXXXX and BACKGR=XXXXXX

When **VIDEOMD** is specified, **FOREGR** and **BACKGR** specify the color to use for the foreground (text) and background. The colors are each specified as a six-digit hexadecimal number, where each two digits can range from 00 to 3F. The three digit-pairs represent the intensities of blue, green, and red, respectively. **Note:** On an EGA display, only the leftmost digit of each pair is significant.

MAPLEPRT=type

The **MAPLEPRT** environment variable specifies the type of hard copy *device* that is the default when **Printer Setup** is selected from the **File** menu of the plot screens. Two examples are **laserjet** or **epson9**. The supported printer device types are listed in the section *Supported Printer Device Types* on page 49.

MAPLEPRF=name

The **MAPLEPRF** environment variable is used to specify the hard copy *file* or device that is the default when **Printer Setup** is selected from the **File** menu of the plot screens. Two examples are **lpt1** for the parallel printer port or the file **plot.out**.

SESSMAX=n

The **SESSMAX** environment variable is used to specify the numbers of lines of a session that are to be retained. At any time, the last n lines of the session will be available for review. The default value is 1000, the minimum is 100, and the maximum is 9999. Each session line takes up about 250 bytes of memory. Thus, the default requires 250 K, the minimum uses 25 K, and the maximum uses 2.5 MB.

Compatibility Considerations

Maple V runs in the 32-bit protected mode of the 80386 or 80486 chip. As a result, it may clash with other programs that do likewise.

The current version of *Maple V* for DOS does run under Microsoft Windows. It is not, however, a Windows application; it is the DOS version running *in* a window. For more information see the chapter *Maple V Release 3 for DOS Running Under Microsoft Windows* on page 57.

There is also a possibility that some *Terminate and Stay Resident* programs (TSRs) will clash with *Maple V*. If you have trouble running *Maple V*, reboot your computer without any TSRs and see whether the problem disappears.

Virtual Memory

Maple V makes use of virtual memory. This means that the complexity of calculations is not limited by how much memory (RAM) you have in your computer. When *Maple V* notices that it is running out of memory, it *pages* pieces of memory to disk to free some RAM. The amount of memory that you have available equals the sum of the available RAM in your machine and the free space on your hard disk. For example, if you have 4 MB of available RAM in your computer and 20 MB of free space on your hard disk, *Maple V* will think that you have 24 MB of memory.

When *Maple V* pages memory to the disk, it creates a temporary file in the root directory. The file is created in the root directory for efficiency reasons. Normally, you need never worry about this file, because *Maple V* removes it before you return to DOS. However, if you abort *Maple V*, or if there is a power failure while *Maple V* is running, the paging file may not be removed. In this case, you must remove it yourself.

The paging file has a name consisting entirely of digits and the letters from A to F. It has no extension. For example, if you type **DIR** in the root directory, you might see a file like this in the list:

```
0A2B42E9              0    2-26-90  11:34am
```

Notice that the size of the file is 0. This occurs because the file has not been properly closed. You should erase any such files using the DOS **ERASE** command. However, the space used by the file is still occupied, because it was never closed. To recover this space, use the DOS **CHKDSK** command with the /F parameter:

CHKDSK   /F

CHKDSK reports that one or more *clusters* were found in one or more *chains*. It will then ask if you wish to convert these to files. Type **N** in response to this question and press the **Enter** key. When **CHKDSK** is finished, all unallocated clusters will be reclaimed.

The root directory is used to store the paging file because it is much faster to access than a subdirectory. If you would rather have *Maple V* use a subdirectory or the root directory of another drive that has more space, use the **CFIG386** utility provided in the **BIN** directory to change the executable to use the desired location. For example:

```
CD   C:\MAPLEV3\BIN
CFIG386   MAPLE.EXE   -SWAPDIR   C:\MAPSWAP
```

or

```
CD   C:\MAPLEV3\BIN
CFIG386   MAPLE.EXE   -SWAPDIR   D:
```

Maple V increases the size of its paging file arbitrarily, depending on how much virtual memory it needs. If you want to limit the size of the paging file, use **CFIG386** with the **-MAXSWFSIZE** option. For example, to limit the paging file to 10 MB:

```
CD   C:\MAPLEV3\BIN
CFIG386   MAPLE.EXE   -MAXSWFSIZE   10485760
```

When operating on a network using a local machine with no hard disk drive but at least 4 MB of memory, you may wish to disable virtual memory entirely to reduce network traffic. To do this, use:

```
CD   C:\MAPLEV3\BIN
CFIG386   MAPLE.EXE   -NOVM
```

Maple V Memory Requirements

Maple V requires 3 MB of extended memory and will use more if your computer has it. In addition, *Maple V* requires 450 K of conventional memory—that is, memory below the 640 K DOS limit—to run its built-in DOS extender, two-dimensional and three-dimensional rendering programs, and the MAPLEDIT text editor. To ensure that enough memory is available for the rendering programs, *Maple V* reserves 320 K of conventional memory when it starts.

If you have a large number of device drivers or TSR programs installed, you may find that *Maple V* refuses to run. There are two solutions to this problem, one of which is to remove any unneeded device drivers and TSRs. If you cannot remove enough drivers and TSRs to allow *Maple V* to run, you can reconfigure *Maple V* to reserve less memory. Do this using the **CFIG386** program described in the previous section, with the **-MINREAL** option. For example:

```
CD   C:\MAPLEV3\BIN
CFIG386   MAPLE.EXE   -MINREAL   4800H
```

This tells *Maple V* to reserve only 288 K of conventional memory. **4800H** is a hexadecimal number indicating how many 16-byte paragraphs of memory to reserve; **4800H** = 18,432 paragraphs = 29,4912 bytes = 288 K. If this change still will not allow *Maple V* to run, try **4000H**—that is, 256 K of memory—instead of **4800H**. Some other values to try are:

5000H	320 K
4000H	256 K
3800H	224 K
3000H	192 K

Using Maple V on a Network

If you are installing *Maple V* on a network, the **BIN** directory should be readable by everyone (all DOS computers on the network) and should be in everyone's **PATH**. The **LIB** directory should be readable by everyone, and a **MAPLELIB** environment variable should be set up so *Maple V* can find the **LIB** directory. Both directories can be read-only.

You may also want to specify where *Maple V* should create its virtual memory paging file and place a limit on the maximum size to which the paging file can grow. It is a good idea to create the paging file on the user's own hard disk so that network traffic is minimized. Each time *Maple V* is started, it creates a paging file, which is removed when *Maple V* terminates. See the section *Virtual Memory* on page 39 for details.

The Status Line

The bottom line of the screen is reserved as a status line displaying information about the current state of your session. Because the same keys, such as the function keys, may have different actions depending on the mode *Maple V* is in, the status line changes to remind you of some of the editing and function keys available in the current mode. For example, in normal session mode, immediately after starting *Maple V*, the status line indicates the total number of bytes allocated by *Maple V* and the number of seconds of CPU time used. To indicate when input/output capture mode has been turned on, an **I** and/or **O** appears within the parentheses on the status line. For more information, see *Input/Output Capture Mode* on page 44.

Editing in Maple V

There are many ways to edit expressions, procedures, and files within *Maple V*. The following sections outline some of these methods, and others are mentioned elsewhere in this chapter.

The Command Line Editor

When entering a *Maple* statement or command, you can use several of the keys on the keyboard to edit what you are typing. The ← and → keys move the cursor left or right one character at a time, without erasing any characters. The **Home** and **End** keys move the cursor to the beginning or end of the line, respectively. The **Backspace** key moves the cursor left and deletes the character there. The **Delete** key deletes the character at the cursor position. The **Insert** key turns insertion mode on or off; insertion mode is initially on. Pressing **Esc** erases the entire line.

Sometimes you will want to edit a line typed earlier. This is most common when you want to make a small change and see what effect it has, or you want to correct a typing error that you did not notice until after pressing **Enter**. *Maple V* maintains a list of the last 100 lines that you typed. The ↑ and ↓ keys scroll through this list. If you want to find a particular line quickly, you can type the first few characters of the line and then press the **F2** key. *Maple V* searches backward through the list for the most recent line that begins with those characters. If no such line is found, the line you typed remains unchanged, and the cursor moves to the beginning of the line.

Expression Editing

The previous section described how you can edit lines that you typed earlier. Using the procedure **xed()**, you can edit any valid *Maple* expression with the *Maple* text editor **MAPLEDIT**. The procedure **xed()** takes a single parameter, which can be either an unevaluated name or an expression.

The parameter is evaluated and converted into line-printed form—that is, the form in which you would type an expression. The **MAPLEDIT** text editor is then invoked to let you edit the expression.

The text editor makes use of the various editing keys ($\leftarrow, \rightarrow, \downarrow, \uparrow$, **Home**, **Insert**, and so on). Pressing **Alt + H** while the text editor is active displays a list of all the keys that you can use to edit.

When you have finished editing the expression, press **F2** to save the changes you made, and then **F3** to exit the text editor and return to *Maple V*. The expression is then read back into the session. If the parameter you passed to **xed()** was an unevaluated name, the expression is also assigned to that name.

Online help for **xed()** is available with the command **?xed**. The help page contains examples that demonstrate the different methods of calling **xed()**.

File Editing

Just as **xed()** lets you edit an expression, the procedures **fed()** and **fred()** let you edit a file. Both of these procedures take a name as the single argument and invoke the text editor, described in the previous section, on the file specified by that name. If the name contains any punctuation characters, it must be enclosed in back-quote characters. The name must be given in *Maple* format rather than DOS format. The difference between *Maple* format and DOS format is that *Maple* uses a forward slash (/) rather than a backslash (\) to separate directory names in the filename path. For compatibility with DOS, each path component should be at most eight characters long. Also, DOS is not case-sensitive; whereas *Maple* is case-sensitive, so the unique *Maple* filenames **MyFile** and **MYFILE** map to the same DOS filenames.

Both **fed()** and **fred()** function the same way except that **fred()** reads the contents of the file into *Maple V* after you have finished editing it. This makes **fred()** very useful during the development of procedures; you can go back and forth between editing and debugging a procedure without having to exit *Maple V*.

Online help for **fred()** and **fed()** is available by entering **?fred** or **?fed** at the input prompt.

Session Review Mode

Maple V automatically saves the last 1000 lines of your session in a session log. This session log can be printed, saved to disk, or viewed with the browser.

Pressing **F5** from the *Maple V* session invokes the browser with a copy of the session log. When you are in the browser, you can highlight a set of lines that can then be edited or read back into the session. The contents of the browser are manipulated exactly as explained in the section *Maple V's Help Facility* on page 54.

Note that the browser contains only a *copy* of the session log. Editing this copy, by pressing **F5** again, changes only the contents of the browser for this use. The next time you invoke session review mode, the browser will contain a new copy of the up-to-date session log.

Input/Output Capture Mode

At any time during the session, you may specify that a copy of the subsequent input and/or output be saved to a file. This feature is accessed from the menu. To start capturing the input and/or output, choose **Begin Capture** from the **Session** submenu. You are asked to enter a filename and to specify with **Y** or **N** whether the input or output (or both) are to be captured. After you have entered the requested information, capture mode is turned on and you are returned to the session. To indicate that capture mode is currently active, the letters **I** and/or **O** appear within parentheses on the status line, depending on whether input and/or output is being captured. To end capture mode, choose **End Capture** from the **Session** submenu. After ending capture mode, the letters **I** and **O** no longer appear on the status line.

Printing the Session Log

To print the current session log or save it to disk, choose **Print Session Log** from the **Session** submenu. Typing **N** in the **Print to File** field sends the session log to your printer. Typing **Y** in the **Print to File** field and a valid DOS filename in the **File Name** field saves the session log to disk.

Preserving Maple's State

At any time during the session, you may preserve the current state of the session for a future use of the program. One reason to save the session's state is that you have to exit the program and wish to continue with your calculations another time. Or, if you reach a certain point in your session where you have two or more directions in which your calculations can go, you may wish to preserve the state of the session at this branch point.

When you choose **Save Session** from the **Session** submenu, you are prompted to supply a filename. The information necessary to recreate the session is saved in two files: a file with a **.m** extension that saves the state of the *Maple* variables and procedures, and a file with a **.ses** extension that saves the current state of the session log and command list.

When you choose **Load Session** from the **Session** submenu, you are prompted to supply a name. This name must be the same one that was used previously to save the session. The information necessary to recreate the session is loaded into *Maple V*. For both **Load Session** and **Save Session**, pressing **F10** at the filename prompt displays the file browser.

Note: Loading a previous session is normally done at the *beginning* of a new session, when very little else has already been defined. By loading a session, you import all the variables and procedures that were defined during the previously saved session. Doing this in the middle of a current session can seriously alter what you have done in the session to date. Also, the command list and session log from the current session will be replaced by the command list and session log from the previously saved session.

Manipulating Graphical Output

Plots are generated with commands such as **plot, plot3d, animate, animate3d, display**, and **display3d**. For explanations and examples of how to use these functions, consult the online help with the commands **?plot, ?plot3d**, and so on. The following sections explain how to display plots on the screen, obtain a printout of a plot, and save a plot to a file.

Two-dimensional Plots

Two-dimensional plots are created principally with the **plot** command. Changes to the appearance of a two-dimensional plot can be controlled through options to the **plot** command or through the menu items available in the two-dimensional plot window. Once you have a two-dimensional plot window on the screen, pressing **F10** allows you access to the two-dimensional plot menus.

Manipulating Graphical Output

Note: If you re-create a two-dimensional plot by re-entering the input line that originally created the plot, the plot displays as originally drawn, without any changes that were made via the two-dimensional plot menus. For each new plot that is created, the values of the options in the two-dimensional plot menus are reset to the defaults.

- The **File** menu of a two-dimensional plot window allows you to **Print** the plot. **Note:** The **Print** option allows you to print to a printer or to a file that can then be sent to a printer or incorporated into a larger document. To print a plot, first select **Printer Setup** and fill in the **Printer Type** and **Device or File Name** fields. The supported printer types are listed in the section *Supported Printer Device Types* on page 49.

- The **Style** menu of a two-dimensional plot window allows you to view the plot in either **Patch**, **Patch w/o Grid**, **Line**, or **Point** mode. The default mode is **Patch** mode. In most cases, **Patch** mode and **Line** mode are the same; however, certain plot commands, such as the **densityplot** command, have a different representation in **Patch** mode versus **Line** mode. The **Style** menu also has submenus for setting **Line Thickness**, **Line Style**, and **Point Style**.

- The **Axes** menu of a two-dimensional plot window allows you to put **Boxed**, **Framed**, or **Normal** axes on the graph, or to have no axes displayed. The default is to have **Normal** axes displayed.

- The **Projection** menu of a two-dimensional plot window allows you to view the plot with **Constrained** or **Unconstrained** scaling. The default is to display the plot **Unconstrained**.

Once you have set all your two-dimensional plot options using the two-dimensional plot menus, **Escape** to the main plot screen, which has the **Enter-Redraw** option, and press **Enter** to redraw the plot with the newly set options.

Three-dimensional Plots

Three-dimensional plots are created principally with the **plot3d** command. Changes to the appearance of a three-dimensional plot can be controlled through parameters to the **plot3d** command or through the menu items available in the three-dimensional plot window.

Note: If you re-create a three-dimensional plot by re-entering the input line that originally created the plot, the plot displays as originally drawn, without any changes that were made via the three-dimensional plot menus. For each new plot that is created, the values of the options in the three-dimensional plot menus are reset to the defaults.

Rotating a Plot

To rotate a three-dimensional plot, press any one of the four arrow keys on the keyboard and a rotation box appears. You can rotate this box using the arrow keys. The arrow keys rotate the plot in increments of 5 degrees, whereas the number keypad with **Num Lock** turned on rotates the plot in increments of 1 degree. When you have the box at the desired angle, press **Enter** to redraw the plot at this new angle.

The 3D Plot Menus

- The **File** menu of a three-dimensional plot window allows you to **Print** the plot. **Note:** The **Print** option allows you to print to a printer or to a file. To print a plot, first select **Printer Setup** and fill in the **Printer Type** and **Device or File Name** fields. The supported printer types are listed in the section *Supported Printer Device Types* on page 49.

- The **Style** menu of a three-dimensional plot window allows you to view the plot in many different styles. The default is to display the plot in **Hidden Line** mode with a grid on the surface; however, you can also display the graph in patch mode with a grid (**Patch**) or without a grid (**Patch w/o Grid**) or patch mode with contours (**Patch and Contour**). Other options in the **Style** menu are **Patch**, **Contours**, **Wire Frame**, and **Point**. The **Style** menu also has submenus for setting **Line Thickness**, **Line Style**, **Point Style**, and **Grid Style**.

- The **Color** menu of a three-dimensional plot window allows you to change the color scheme of the graph and add lighting to the graph. **Light Schemes 1**, **2**, **3**, and **4** are predefined light schemes set for your convenience. If you specify your own light scheme on the command line while creating the three-dimensional plot, then the **User Lighting** option in the **Color** menu becomes available so that you can toggle between **No Lighting**, **User Lighting**, and **Light Scheme 1**, **2**, **3**, or **4**.

- The **Axes** menu of a three-dimensional plot window allows you to put **Boxed**, **Framed**, or **Normal** axes on the graph, or to have no axes displayed. The default is to have no axes displayed.

Manipulating Graphical Output

- The **Projection** menu of a three-dimensional plot window allows you to view the plot with **Constrained** or **Unconstrained** scaling. The default is to display the plot **Unconstrained**. The **Projection** menu also allows you to view the plot from a **Near**, **Medium**, **Far**, or **No Perspective**. The default is to display the graph with **No Perspective**.

Once you have set your three-dimensional plot options using the three-dimensional plot menus, **Escape** to the main screen (the screen with no menu) and press **Enter** to redraw the plot with the newly set options.

Animation

An animation window is a variation of a two-dimensional or three-dimensional plot window. When you create an animation using the **animate** or **animate3d** commands, or **display** or **display3d** commands with the option **insequence=true**, a window with multiple frames is drawn. These are the frames of the animation; the animation itself runs in the top left frame.

In both a two-dimensional and a three-dimensional animation window, you get the standard **F10-Menu** option giving you access to all the standard two-dimensional and three-dimensional plot menus. You also have animation control as follows:

F4	**Direction**—allows you to play the animation in either forward or reverse mode
F5	**Fast** (speed up)
F6	**Slow** (slow down)

In forward mode, the animation plays through as entered from the command line. In reverse mode, the animation plays backward.

Printing and Saving Plot Output

There are two methods of sending plot output to a printer or file. The simplest method is to choose the **File** item from the menu in the display driver. The second method is to specify that the **plot** or **plot3d** command that generates the plot send the plot output directly to the printer or to a file without first displaying it.

To print a plot directly, specify the device type and output file. The following code prints a plot on a 9-pin EPSON printer via parallel port number one. Substitute the appropriate device types and output file for your configuration.

interface(plotdevice=epson9, plotoutput=lpt1);
plot3d(x^2 + y^2 , x = -5..5, y = -1..3);
interface(plotdevice = ibm);

The last command tells *Maple* to display future plots on the screen. For further information, consult the online help for the **interface** command.

Supported Printer Device Types

Maple V can generate graphic output for the following printer devices and file formats. Use the abbreviations **epson24**, **epson9**, **epson9hi**, and so on when specifying the selection.

Printers

canon	Canon bubble jet
cps	color PostScript printer
deskjet	HP Deskjet and PCL4 laser printers
epson24	EPSON-compatible 24-pin dot matrix printers
epson9	EPSON-compatible 9-pin dot matrix printers
epson9hi	EPSON-compatible 9-pin printer in hi-res mode
hp7470	HP 7470 plotter
hp7475	HP 7475 plotter
hp7550	HP 7550 plotter
hp7585	HP 7585 plotter
i300	Imagen 300 laser printer
ibmpro	IBM Proprinter X24 dot matrix printer
ibmquiet	IBM Quietwriter dot matrix printer
laserjet	HP LaserJet printer
ln03	DEC LN03 laser printer
oki92	Okidata Microline92 dot matrix printer
paintjet	HP PaintJet printer
postscript	PostScript printer
toshiba	24 pin Toshiba printers

File Formats

bmp	Windows bitmap file

Using the Maple V Menus

gif	GIF file format
hpgl	generic HP plotter format
pcx	Z-soft PCX file format
pic	UNIX troff *pic* format
unix	UNIX plot utility format

Using Maple V Plots in Other Programs

Some software applications allow you to import graphical objects stored as PostScript or HPGL files. To create a PostScript or HPGL file containing a *Maple V* plot, specify the correct device type (for example, **postscript** or **hpgl**) and an output file (for example, **mypict.ps** or **mypict.hp**) when using one of methods of saving plots explained in the section *Printing and Saving Plot Output* on page 48.

Using the Maple V Menus

For the beginner, the menu system provides a simple interface to some of *Maple*'s key functions. For the more experienced user, the menu offers a method of customizing *Maple V*, eliminating the need to retype commonly repeated or difficult-to-remember command sequences. Modifications to the menu are made from within the menu of the session. Thus, it is no more difficult to create a custom menu configuration than it is to use the menu itself.

To activate the menu, press **F10**. A **Menu Bar** appears at the top of the screen with a list of items to choose from.

Items are chosen either by highlighting the item with the arrow keys and pressing **Enter**, or by pressing the letter key corresponding to the highlighted letter in the desired item. Invoking one of the top-level menu items usually causes a submenu to be displayed.

Items in submenus are chosen in exactly the same way as top-level items. Pressing **Esc** while a submenu is displayed returns you to the top-level menu. To exit the top-level menu, press **Esc** again.

The **Session** item in the **Menu Bar** is the only item that is common to all menu configurations. *All* other items are definable by the user. The simplest **Menu Bar** would be one that contains only **Session**. The **Session** submenu includes the following items:

- **Load Session..** and **Save Session..** are explained in the section *Preserving Maple's State* on page 44.
- **Load Menu..**, **Save Menu..**, and **Modify Menu** create new menu configurations. To load or save a menu, choose **Load Menu..** or **Save Menu..** and enter the filename used to store the menu configuration. Menu configurations are stored to disk with the filename extension **.MNU**. You do not need to type the extension; it is supplied automatically. It is a good idea to save the menu before you modify it, in case you want to restore it later.
- **Begin Capture..**, **End Capture**, and **Print Session Log..** are explained in the section *Input/Output Capture Mode* on page 44.
- **Quit** exits the *Maple V* session.

When *Maple V* is started, it loads in the default menu configuration. The default configuration is stored in the file **MAPLE.MNU** in *Maple*'s library directory. If you would rather use your own menu configuration as the default, save **MAPLE.MNU** under a different name as a backup, and save your menu with the name **MAPLE.MNU**.

Note: For the Student Edition, the menu configuration is stored in the file **STUDENT.MNU** instead.

There are three levels involved in modifying the menu. The first level is invoked by choosing the **Modify Menu** option in the **Session** submenu. The first-level window is a window from which you select the item in the **Menu Bar** to modify. The second level is activated after you make your selection. The second-level window contains a list of the submenu items corresponding to the first-level item you selected. After choosing a submenu item in level two, a large window will open, allowing you to modify the definition of the submenu item. The following sections give a further explanation of each level.

Level One

Invoking the **Modify Menu** option opens a window with space for exactly 16 entries. Each of the current first-level menu items, except for the **Session** item, will appear in

Using the Maple V Menus

one of the 16 slots. Choose the first-level menu item that you want to modify or choose an empty slot to create a new first-level item. After you make your choice, a second window opens and you are in the second level.

Level Two

The second-level window has room for exactly 16 entries. Each of the current submenu items for the selected first-level menu item appears in one of the slots. At this point, there are three operations you may perform.

1. Select **Edit Menu Name** to change the name of the first-level menu item you selected previously. Setting the name to blank deletes that item from the first-level menu or **Menu Bar**.

2. Select one of the submenu items to modify, or choose a blank slot if you want to add a new item.

3. Press **Esc** to return to level one.

After selecting a submenu item to modify, you are in level three.

Level Three

A large window opens that lets you edit the *definition* of the submenu item chosen in the second level. If you chose a blank second-level slot, you can now enter the definition for a new submenu item. The definition can be a maximum of 16 lines long. As shown on the status line, the ↑, ↓, ←, and → keys are used to edit and move to different fields in the third-level window.

After you have completed modifying the item, press **F3** to accept the new definition. To delete the item, set the name of the item to blank and press **F3**.

There are two types of entries in the definition field lines: *Maple* commands, which are to be read into the session when this item is chosen, and lines beginning with %. Lines beginning with % are used to obtain input from the user before the *Maple* commands are read into the session. All lines beginning with % must come before any *Maple* commands.

Lines beginning with %n (where n is a digit from 0 to 9) designate that the user be asked for input when the menu item is invoked. The text following the %n is used to prompt the user for input. When the user activates an item with the % notation in its definition, a dialog box opens and requests the user to enter expressions into each of the % fields. Then all occurrences of % variables (not to be confused with % labels in *Maple* output; see **?interface**) in the code of the item definition are replaced with the user's typed input.

Executing *Maple* code by invoking an item from the menu does not echo the commands to the screen. To cause commands to be echoed, end the line of code in the menu definition with a tilde (˜). Only those lines ending with a tilde are echoed. This feature is useful to prevent cluttering up the session.

A Sample Menu Entry

Here is an example of a menu entry from the default menu (**MAPLE.MNU**) supplied with the *Maple V* distribution:

```
                              DOS Window
     Operations    Packages    Translate    Edit    Interface    Session

 Command Name: [Sum..         ]
                              Maple Commands
 [%1Find sum over                                                        ]
 [%2 ranging from                                                        ]
 [%3            to                                                       ]
 [sum(",%1=%2 .. %3);~                                                   ]
 [                                                                       ]
 [                                                                       ]
 [                                                                       ]
 [                                                                       ]
 [                                                                       ]
 [                                                                       ]
 [                                                                       ]
 [                                                                       ]
 [                                                                       ]
 [                                                                       ]
                   ←,→,↑,↓-Move  Enter-Next  F3-Done  Esc-Cancel
```

When this menu entry is invoked, by pressing **F10, O,** and **U**, a dialog box appears. If you then fill in the fields like this:

```
 Find sum over: [i                                                       ]
    ranging from: [0                                                     ]
              to: [100                                                   ]
```

the following command is entered into your session and executed (and echoed because of the ˜):

sum(" , i = 0..100);

Getting Started

This will compute the sum of the previous result with the summation variable *i* ranging from 0 to 100.

Maple V's Help Facility

The usual method of accessing help in *Maple V* is to use the **?** syntax. At the input prompt, type **?** and then the topic for which you want help. For example, typing **?fred** loads a copy of the help page for the procedure **fred()** into the browser. If you type something after the **?** that *Maple V* does not recognize, alternative topics are suggested.

The second method of bringing up a help page is the **F1** key. Pressing **F1** displays an interactive topic browser. *Maple*'s help pages are organized here in a logical hierarchy. A topic with a pointer ▷ has related subtopics beneath it. The list on the far left is the highest level of the hierarchy and contains several general topics—for example, **Graphics**, **Mathematics**. Each list to the right is one level lower.

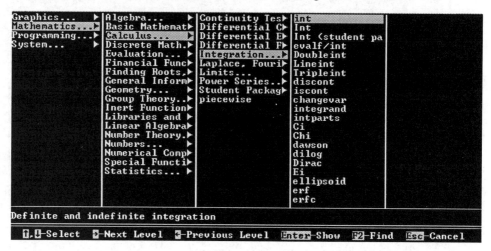

While in any topic list, use the arrow keys to scroll up and down the list and to move to lists on the left or right.

Highlighting any topic updates the synopsis field, located near the bottom of the topic browser, with a brief description of the help page for that topic. To open a help page from the topic browser, press **Enter** while the topic is highlighted. The topic browser is particularly useful when you are not sure of the exact name of the help page you are interested in viewing.

Another feature available within the topic browser is keyword search, **F2-Find**. When this option is selected, you will be prompted to supply a string of characters—for example, *plot*. A list of all the help pages whose synopsis contains that string is given, from which you can select the page you wish to view.

After choosing a help topic using any of these methods, you will be in the help page viewer.

Use the ↓, ↑, **Page Up**, **Page Dn**, **Home**, and **End** keys to view various parts of the help page. When you are in the viewer you can highlight a set of lines that can then be edited or read into the session. To highlight a set of lines, go to the start of the set you want to highlight and press the **Space Bar**. Then scroll down to the last line in the set and press **Space Bar** again. You can then perform two operations on the highlighted lines:

1. Press the **Enter** key to quit the viewer and read the highlighted lines into the session.

2. Press **F5** to invoke the *Maple* text editor with the highlighted lines. Edit the contents of the text editor. If you want to replace the contents of the viewer with the contents of the editor, press **F2** (save) and **F3** (quit) in succession. You will be returned to the viewer with the new contents. **Note:** The entire contents of the viewer are now highlighted, allowing you to quit the viewer and read the viewer's contents into the session simply by pressing **Enter**. Pressing **Esc** will exit the viewer without reading anything into *Maple*.

Note: The viewer contains only a *copy* of the help page. Editing this copy changes only the contents of the page for this use. The next time you access help for the same subject, the viewer will contain the original help page.

Maple V Release 3 for DOS Running Under Windows

It is also possible to run a copy of the DOS version of *Maple V Release 3* within a Windows window. There are several differences in the operation of *Maple V* under this scenario, and they are detailed in the following sections.

Setting Up Maple V under Windows

Verify that *Maple V Release 3* has been installed as per the instructions in the chapter *Installing Maple V Release 3* on page 3, and that all the required *Maple V Release 3* Windows icons have been prepared.

Running Maple V under Windows

Bring the **Maple V Release 3** group window to the top by clicking it. Double-click the **Maple V Release 3 - DOS** icon to launch *Maple V*. A window will open, and the DOS version of *Maple V* will start in it. **Note:** Because of the way *Maple V Release 3* uses the screen, Microsoft Windows will probably prompt you to run *Maple V Release 3* in full screen mode. You should do so.

Special Features for Windows

Maple V Release 3 for DOS has several special features that are available only under Microsoft Windows, in **386 enhanced** mode. These are described in the following sections.

Copy and Paste

During a *Maple V* session, the **F6** key can be used to copy or paste text. In *Maple*'s standard interactive mode, pressing **F6** pastes any textual contents of the Clipboard into the session, as if you had typed it at the command line and entered it. A semicolon and **Enter** are automatically supplied. This lets you create files of *Maple* commands,

Special Features for Windows

perhaps in Windows Write, for selective insertion and execution in a session.

In **Review** mode (having pressed **F5**), pressing **F6** copies the selected text onto the Clipboard, much as pressing **Enter** causes the text to be executed. Recall that lines of text are selected by pressing the **Space Bar**. This feature lets you transfer parts of your session to other Windows applications, such as Write.

Multiple Maple V Sessions

Windows is a multitasking operating system and can therefore support multiple simultaneous *Maple V* sessions. To start another session while one is already running, double-click again the **Maple V Release 3 - DOS** icon.

Keep in mind that these sessions are *entirely independent.* Anything you do in one does not affect another, unless you are sharing input/output files between them. You *can* transfer *Maple* commands from one to another using the copy and paste facilities described in the previous section.

If you try to generate an EGA or VGA plot when *Maple V* is running in a window, Windows displays a box informing you that the application—in this case *Maple V*—has been suspended until you switch to full screen mode.

Other Applications

Mint: The Syntax Checker

The Mint application provides a robust syntax-checking facility for files of *Maple* code. Mint can be accessed from within Windows or from the DOS command line as follows:

- Windows: Click on the **Mint** application icon.
- DOS: Enter **mint** at the command line prompt.

Within Mint, you are prompted for the name of the file you want to pass through the syntax checker. Mint also prompts you for the level of information you wish to see. Respond with one of the following:

0	No information
1	Severe error messages only
2	Severe and serious error messages
3	Warnings and error messages
4	Usage, warning, and error messages

Once Mint finishes processing your input file, the output it generates is presented to you within a Mapledit session. This lets you browse through Mint's comments. If you want to save the comments in a file, press **F7**, type the name of the file, press **Enter**, and then press **F2** to save the file. Press **F3** to exit Mapledit.

For more information on the workings of Mint, refer the the online help page for **mint**.

March: The Maple Archive Manager

The March application provides a manager for *Maple* library archives. The file **maple.lib**, which is in your *Maple* library directory, is an example of a *Maple* archive

March: The Maple Archive Manager

file. March can be accessed from the DOS command line, and its available options are as follows:

- Add **.m** files to an archive:

    march -a *archive_dir filename indexname*

    The **.m** file *filename* is added under index name *indexname* to the archive in the directory *archive_dir*. Multiple pairs of filenames and index names can be specified.

- Update **.m** files in an archive:

    march -u *archive_dir filename indexname*

    The existing entry under index name *indexname* in the archive in directory *archive_dir* is updated to the new **.m** file *filename*. Multiple pairs of filenames and index names can be specified.

- Pack the files in an archive so that they take as little disk space as possible:

    march -p *archive_dir*

- List the content of an archive:

    march -l *archive_dir*

- Create a new library archive:

    march -c *archive_dir table_size*

    The archive is created in *archive_dir*, and the value of *table_size* should be approximately how many files you expect to be in the archive.

- Extract a file from an archive:

    march -x *archive_dir indexname filename*

    This creates a **.m** file called *filename* that is pulled from the index entry at *indexname*.

- Delete a file from an archive:

    march -d *archive_dir indexname*

Only one of these options can be used in a single use of March.

If you make mistakes, appropriate error messages are displayed. For more information, see the help page for **march**.

Other Applications

Mapledit: The Maple Editor

The Mapledit application provides a handy editor for *Maple* files and expressions. Mapledit can be accessed from within *Maple* (for DOS), from within Windows, or from the DOS command line as follows:

- *Maple*: Enter one of the following commands within *Maple* for DOS:

    fred(`*filename*`**)**

    fed(`*filename*`**)**

    where ` is the back-quote character. Both commands will read the file *filename* into Mapledit. If there is no such file available, a new file is created. **fred** also automatically reads *filename* back into the *Maple* session when editing is complete.

- Windows: Double-click the **Mapledit** application icon. You will be prompted for the name of a file to edit.

- DOS: Enter **mapledit** *filename* at the command line prompt.

Within Mapledit, press **F1** to bring up help pages to explain all the available options. To save the work you have done, press **F2**. To exit Mapledit, press **F3**.

m2src

The **m2src** application converts **.m** (binary) files from *Maple V Release 2* into *Maple V Release 2* source files. After a file is converted, it can then be run through **updtsrc**, which is explained in the next section. **m2src** can be accessed from the DOS command line and its available options are as follows:

- Write the results to an output text file *inputfile.t* rather than to standard output:

    m2src -t *inputfile.m*

- Write library procedures in full to an output text file:

    m2src -p -t *inputfile.m*

    This option is necessary if your input file contains *Maple V Release 2* library procedures.

For more information on **m2src** and its available options, see the help page for **m2src**.

updtsrc

The **updtsrc** application converts *Maple V Release 2* source files so that the handling of undeclared variables in *Maple* procedures conforms to the new rules for *Release 3*.

- To read the text file *inputfile.t* and create a source file *outputfile* enter this at the DOS prompt:

    updtsrc *inputfile.t > outputfile*

For more information on the new rules, see the section *Global Variables* in the *Maple V Release 3 Notes*, or refer to the help page accessed with **?updates,v5.3**. For more information on **updtsrc** and its available options, see the help page for **updtsrc**.

Multiple Library Access

Maple V Release 3 supports multiple libraries to make code development a more easily controlled task. Multiple libraries are specified in Windows and DOS by using the **-b** switch when starting a *Maple V* session. The command used is similar to:

maple -b /maplev3/lib -b /maplev3/myself/mydirectory

where the user's own libary has been stored in **/maplev3/myself/mydirectory**. Such a command is used either in Windows in the **File, Properties, Command Line** option of the **Program Manager** or at the DOS prompt when starting a *Maple V* for DOS session.

Index

.m extension
    DOS, 45
    Windows, 24
.ms extension, 24
.ses extension, 45
? syntax
    DOS, 54
    Windows, 29

animation
    DOS, 48
    Windows, 22

command line editing, 42
compatibility, 38
coprocessors, 4
Copy, 25
Cut, 25

Delete, 25
done, 7

Edit menu, 24
editing, 11, 42
    command lines, 42
    expressions, 42
    files, 43
environment variables, 35
exiting *Maple V*
    DOS, 33
    Windows, 7

File menu, 23
fonts, manipulation, 25
Format menu, 25

graphics regions, 14
hardware
    memory size, 8
    requirements, 3
help
    context sensitive, 28
    files, 28
    interface, 28
    keyword search, 28, 55
    with **?**, 29
help browser
    DOS, 54
    Windows, 28
Help menu, 27

input regions, 13
installation, 3
    directory, 4
    procedure, 5
interface
    window types, 10
internal state, 9
    saving, 9
interrupting *Maple V*
    DOS, 33
    Windows, 8

kernel, 8

LaTeX format, 24
libraries, multiple, 62
library, directory, 5

m2src application, 61
MAPLE.INI file, 8

INDEX

Mapledit application, 61
March application, 59
memory, size, 8
menus
    customizing, 51
    DOS, 50
    Edit, 24
    File, 23
    Format, 25
    Help, 27
    loading, 51
    Options, 27
    View, 26
    Windows, 23
Mint application, 59
monitors supported, 37

networks
    installing *Maple V* on, 41

Options menu, 27
output
    replace mode, 27
    size, 26
output regions, 14

page breaks, 25
paging files, 40
Paste, 25
plots
    (Un)Constrained, 19
    animation, 22, 48
    axes, 19, 21, 46, 47
    color, 21
    copying, 14, 18
    creation, 17
    dithering, 21
    lighting, 47
    manipulating, 17
    options, 19, 21, 46, 47
    printing, 24, 48
    redrawing, 27, 46, 48
    rotating, 20, 47
    saving, 48
    three-dimensional, 46
    tool bar, 19, 22
    two-dimensional, 18, 20, 45
    windows, 18
printers
    default, 38
    supported, 49
printing, 24
    page margins, 24
    set-up, 24
problems, 3
prompts, 13, 26

quit, 7

recalculating, 16
regions, 11, 12
    continuous mode, 27
    converting, 11, 25
    graphics, 14
    input, 13
    manipulating, 14, 25
    output, 14
    region groups, 14
    removal, 25
    separator lines, 26
    text, 13
Return, 13

saving
    before exiting, 7
    internal state, 9
    LaTeX format, 24
    option settings, 23
screen color scheme, 36
separator lines, 14, 26

INDEX

session log
    capturing, 44
    printing, 44
Shift + Return, 13
starting *Maple V*
    DOS, 32
    Windows, 7
status bar, 12, 26
    three-dimensional plots, 22
    two-dimensional plots, 19
    worksheets, 12
stop, 7

technical support, 3
temporary files, 5
text regions, 13
tool bar, 26
    functions, 12, 19
        three-dimensional plots, 22
        two-dimensional plots, 19

TSRs, 39

updtsrc application, 62

View menu, 26
virtual memory, 39

Windows
    icons, 6
    set-up, 6
worksheets
    converting, 10, 16
    editing, 11, 15
    fonts, 25
    opening, 23
    opening new, 23
    printing, 24
    recalculating, 16
    recalculation, 10

regions, 11
Release 2, 10
Release 3, 10
saving, 23
that lie, 15
working with, 11